SPOTLIGHT ON NATURE

COUGAR

NELL MUSOLF

CREATIVE EDUCATION · CREATIVE PAPERBACKS

Published by Creative Education and Creative Paperbacks
P.O. Box 227, Mankato, Minnesota 56002
Creative Education and Creative Paperbacks are imprints
of The Creative Company
www.thecreativecompany.us

Design and production by Blue Design, Inc.
Art direction by Tom Morgan
Edited by Grace Beltowski

Photographs by Alamy Stock Photo/Corbis Flirt, 14; Corbis/Marianne Rosenstiehl/Sygma, 29; Dreamstime/Outdoorsman, 10, 28, Seread, 17, 29, Steven Blandin, 18, Sunheyy, 27, 29; iStock/Evgeny555, 26, James Michael Kruger, 11; Getty Images/Alex Hibbert, 9, Alfredo Santamaria / 500px, 21, Easy_Company, 8, 10, 14, 16, 20, 22, Fuse, 21, Ibrahim Suha Derbent, 15, Kathleen Reeder Wildlife Photography, 6, McDonald Wildlife Photography Inc., 4–5, Olena Ruban, 3, 6; Pexels/Pixabay, cover, 1; Shutterstock/creativex, 23, Geoffrey Kuchera, 24, jo Crebbin, 16; Wikimedia Commons/Shahzaib Damn Cruze, 12

Library of Congress Cataloging-in-Publication Data
Names: Musolf, Nell, author.
Title: Cougar / Nell Musolf.
Description: Mankato, Minnesota : Creative Education / Creative Paperbacks, [2026] | Series: Spotlight on nature | Includes bibliographical references and index. | Audience: Ages 10-13 | Audience: Grades 4-6 | Summary: "Leap into the cougar's world (also called mountain lion or puma) with this nature title for middle-grade wildlife lovers. Informational text pairs with a narrative about a single cougar family to spotlight the wild animal's life cycle, supported by infographics and a timeline of developmental milestones"— Provided by publisher.
Identifiers: LCCN 2024048399 (print) | LCCN 2024048400 (ebook) | ISBN 9798889896197 (library binding) | ISBN 9781682777855 (paperback) | ISBN 9798889896999 (ebook)
Subjects: LCSH: Cougar—Juvenile literature.
Classification: LCC QL737.C23 M928 2026 (print) | LCC QL737.C23 (ebook) | DDC 599.75/24—dc23/eng/20241220
LC record available at https://lccn.loc.gov/2024048399
LC ebook record available at https://lccn.loc.gov/2024048400

Printed in India

CONTENTS

MEET THE FAMILY

NORTH AMERICAN COUGARS of Vancouver Island

It's a chilly autumn day on Vancouver Island. The island is located off the coast of British Columbia in western Canada. It is home to its own **subspecies** of cougar, *Felis concolor vancouverensis*. Between 600 and 800 cougars live on Vancouver Island. Most residents will never see one since the large mountain cat is quick, sly, and prefers to live by itself.

On this fall day, a female North American cougar sits in an isolated cave. The cave is in a deeply wooded forest, protected from the wind. The cougar has spent the previous weeks getting the den ready for her offspring. The walls are lined with moss, making the space cozy. The cougar will give birth alone. Her litter could be anywhere from one to six cubs. One thing is certain: by the end of the day, she will be a mother.

CLOSE-UP

Spots

Cougars are born with dark brown spots. Their tails have dark rings. As a cougar grows, the spots fade and its coat becomes a **uniform** shade of tan. Seeing a cougar with spots means it is still a cub and hasn't yet reached adulthood.

CHAPTER ONE

LIFE BEGINS

Cougars—also known as mountain lions, American lions, pumas, and panthers—live throughout North America and South America. Most cougars live in the western half of North America from Canada down to Mexico, but there are some found as far south as Chile. There is also a small population of cougars in southwest Florida. Cougars live in forests, mountain regions, deserts, and canyons. Wherever they live, cougars **adapt** to the terrain.

As members of the big cat family, adult cougars can measure up to 9 feet (2.7 meters) long from nose to tail. Full-grown males weigh around 180 pounds (81.6 kilograms). Females weigh in at 120 pounds (54.4 kg). Cougars are born with dark spots that fade by six months. Adults have solid tan coats. They have round heads and erect ears. They have excellent eyesight and hearing, but they have a weak sense of smell. Cougars are known for their strength. They have long, lean, muscular bodies. They are

NORTH AMERICAN COUGAR MILESTONES

DAY 1

- Born
- Eyes tightly shut
- Tan-colored coat has dark spots
- Weight: 12 ounces (340 grams)

built to move stealthily as they hunt for prey. Their front feet are larger than their back feet. They have four **retractable** claws on their hind paws and five on their forepaws with one **dewclaw**. Their claws adjust for clutching prey. Their long tails help them balance their bodies.

Cougars are **carnivores**. Their diet consists of mammals such as deer, moose, bighorn sheep, and goats. If larger prey isn't available, they eat small mammals including rodents, birds, and domestic animals such as dogs and cats. They prefer to hunt at night. With their excellent eyesight, they can spot prey easily. They will travel far distances in search of food. Cougars are fast and agile. They can run up to 50 miles (80.5 kilometers) per hour. They are **ambush** predators. Young

CLOSE-UP

Color-changing eyes

At birth, a cougar's eyes are tightly shut. They are blue when they first open 5 to 19 days later. By 16 months, the cougar's eye color will change to a golden green and will not change again for the rest of its life.

FEATURED FAMILY

Welcome to the World

Deep in the woods on Vancouver Island, the female cougar is lying in the moss-lined den she has made for herself. After three months of being pregnant, her cubs are about to be born. One cub emerges, then another, making for just two in this litter. Their eyes are tightly shut, and their bodies are covered with brown spots. They cannot hear. The cubs give weak mews before beginning to nurse. They will be completely dependent on their mother for the next few weeks. Mother and cubs fall into a deep sleep.

and sick cougars might be attacked by other cougars, bears, or wolves. Otherwise, humans are cougars' only predators.

While adult cougars are known for their strength and excellent hunting skills, they begin life like all mammals. They are weak and completely dependent upon their mothers. Newborn cubs are blind. They are fed by their mother's milk. As cubs grow, they communicate with their mothers through whistles, chirps, and mews. They will remain with their mothers for the first few years of life.

DAY 8

- Eyes are fully open and blue
- Continues nursing
- Fully dependent on mother

CLOSE-UP

Good grip

A dewclaw is an extra digit on a mammal's paws. The dewclaw on a cougar is on its front paws and helps the cougar grip its prey more easily. Cougar tracks will not show the dewclaw unless the cougar is running.

FEATURED FAMILY

First Meal

The newborn cubs are weak. Their sense of smell will never be strong, but they can smell the mother's milk that will keep them alive. One nuzzles against his mother, his tiny mouth searching for her teat. Mother's milk will sustain him until he's old enough to eat meat. Newborns depend on their mother's hunting ability to stay alive. As the cubs get older, their mother will introduce them to meat from birds, elk, and other animals. Then she will teach them how to hunt for themselves.

Cougars can ADAPT to all kinds of environments and challenges.

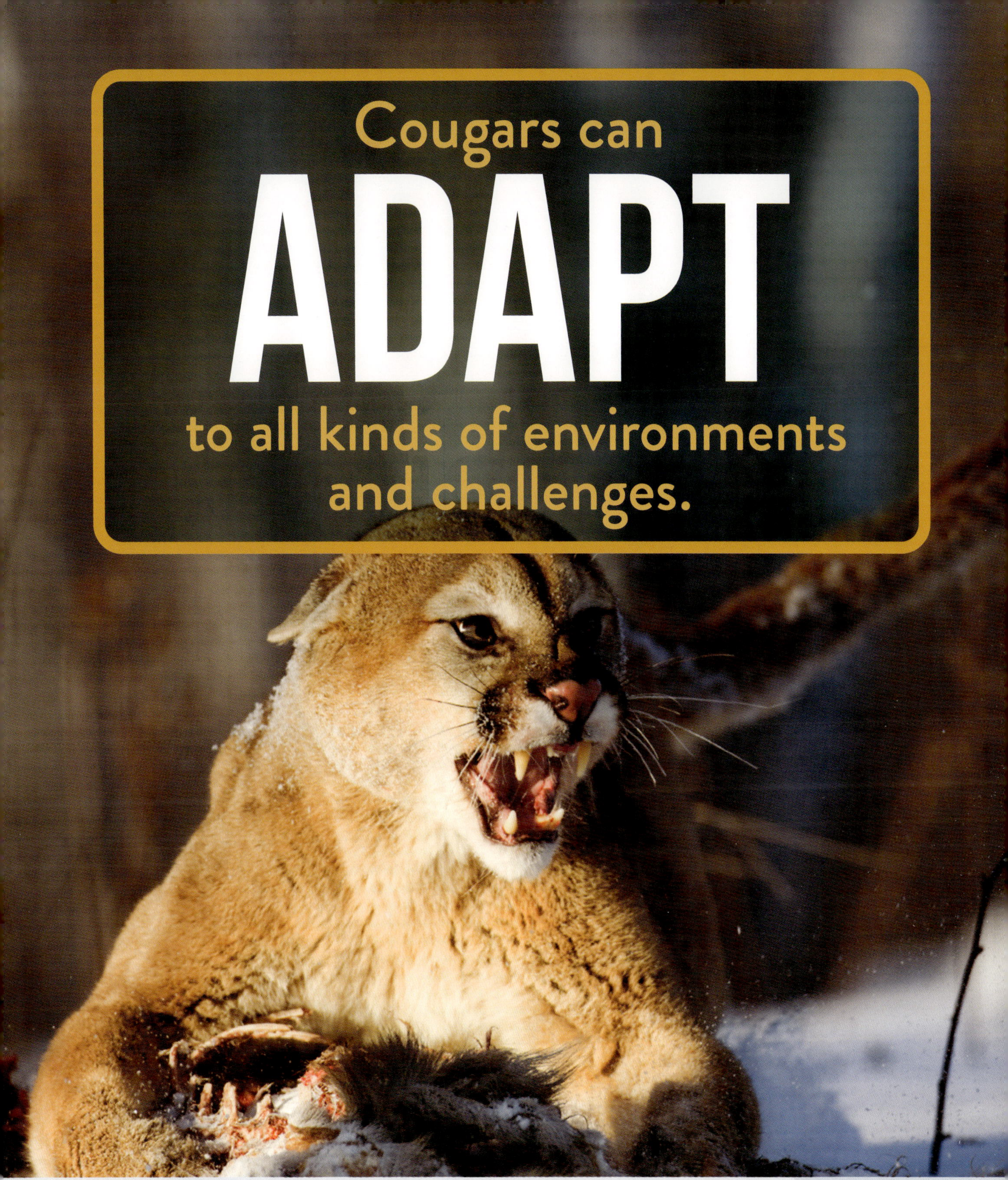

6 WEEKS

- Able to hear
- Canine teeth appear
- Spots on coat still visible
- Starts eating meat

CLOSE-UP

Eyelids

Cougars have black lining underneath their eyes. This makes it look as if they are wearing eyeliner, but in reality it is eyelid skin.

CHAPTER TWO

EARLY ADVENTURES

Cougar cubs stay close to their mothers for 18 to 24 months after being born. The most important lesson cubs learn during this time is how to hunt. Being able to catch their own food is the one skill they can't live without.

Mother cougars begin teaching their cubs to hunt when they are around six months old. They start with smaller prey such as birds, squirrels, and rabbits. Cubs often learn how to hunt using elk calves and fawns as targets. As cougars get bigger and stronger, they can attack and kill full-grown elk.

By two years of age, cougars can hunt on their own. Deer become their prey of choice. Cougars prefer mule deer over whitetail. Mule deer live in open areas, making it easier for cougars to ambush them.

Cougars catch their prey by attacking from behind or from the side. They grab their prey by the neck and pull it to the ground. If they don't break the prey's neck, they will suffocate it. Once the prey is caught, the cougar will eat only a little bit of the meat. Then it buries what's left. After a few hours,

6 MONTHS

- Spots on coat start to fade
- Stops nursing
- Mother begins teaching cubs how to hunt
- Weight: 35–45 pounds (15.9–20.4 kg)

CLOSE-UP

You can't see me...

Cougars hide from other animals and humans by climbing trees, leaping onto ledges, or camouflaging themselves. Often the only sign a cougar has been in an area are tracks and tail drag marks in the mud or snow.

FEATURED FAMILY

Eat Up!

The young cougar cubs love to eat meat. Their mother has been providing it, but the time has come for the cougars to begin hunting for themselves. Their mother leads them to a spot where a rabbit is eating grass. She shows them how to stealthily approach the rabbit from behind. One swift leap and the rabbit is dead. Over the next few weeks, she teaches her offspring how to catch and kill more rabbits, squirrels, and birds. Eventually she teaches them a much bigger kill: deer.

the cougar returns to the kill, which has now cooled. The cougar stays near its kill for the next few days or until it has been fully eaten.

Cougars learn how to take care of themselves quickly. They begin to move farther from their mothers. Siblings stay together for a few months after leaving their mother, but eventually they all become independent.

(16) MONTHS

- Becomes more independent
- Spots continue to fade
- Eye color changes to golden green

CLOSE-UP

...but I see you

Cougars have excellent vision, especially in the dark. They use their amazing eyesight to silently stalk prey and then suddenly leap on them. They can jump as high as 15 feet (4.6 m) into the air before landing on their unsuspecting prey.

FEATURED FAMILY

Give It a Try

The mother cougar has enjoyed having her cubs with her for the past two years. But the day has come for them to leave. All cougars have their own range. Male cougars have ranges that are twice as large as females' ranges. The mother cougar must drive her offspring away from her range. It won't be easy, but she knows it must be done. Patiently, she shoos them farther and farther away from home. The two cubs will stay together for a few months. Eventually, they'll go off on their own.

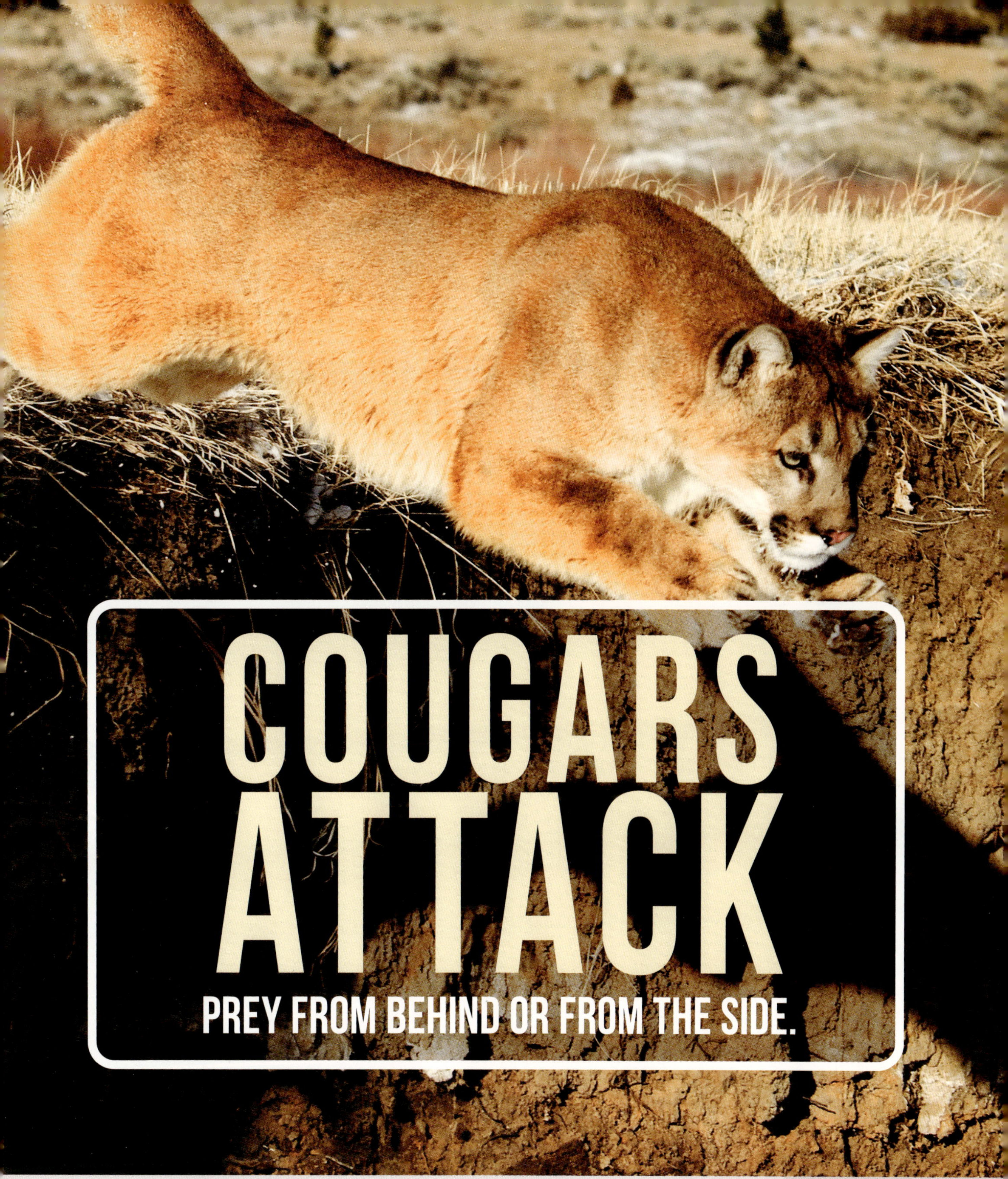

22 MONTHS

- Able to hunt alone
- Leaves mother
- Finds its own range with sibling

CLOSE-UP

Big teeth

Since cougars eat only meat, their teeth are large and sharp. Cougars have special molars that help rip through fur and tear flesh. Their strong teeth along with their strong jaws help with hunting and killing prey.

CHAPTER THREE

LIFE LESSONS

Cougars are **territorial**. They live within a set range of land between 50 and 150 square miles (129 to 388 square kilometers) for males and up to 60 square miles (155 sq km) for females. This is where they find food, water, and shelter. Once cougars separate from their mother and any siblings, they find their own range. They do not live in packs or invite other cougars to live within their range. They mark their territories by leaving piles of leaves, pine needles, or dirt covered with urine and feces.

Mating among cougars takes place between March and June of each year. After mating, male cougars leave parenting to the females and have nothing to do with their offspring. Females can have their first litter when they are between two and three years old. They spend approximately 75 percent of their life either pregnant or taking care of their young. It is possible for females to become pregnant by multiple males, so a single litter can have more than one father. Mother cougars are known to be

2 YEARS

- Cubs separate and live alone
- Females reach sexual maturity and have first litter

very protective of their young and will defend them against predators, sometimes to death.

Other than defending their offspring, cougars are the least aggressive of the world's large cats. They are **elusive** and do not like being seen. They also help the environment by preying on deer and elk that can damage forests through **over-browsing**.

Cougars live between 8 and 13 years in the wild. As with all animals, they face a variety of threats. Hunting for sport is a constant threat to cougar populations throughout North and South America. Another is the **encroachment** of cities into spaces occupied by cougars. As land is developed, cougars are pushed out of their ranges and forced to find new homes.

FEATURED FAMILY

This Is How It's Done

The cougar siblings are hungry. They learned from their mother how to make small kills. Now they are ready for something bigger. Patiently, they wait for a deer to wander into their range. Moving silently but swiftly, they travel throughout the area, their eyes seeking their next meal. Finally, their patience is rewarded. A deer is nibbling on some grass underneath a pine tree. Moments later, the deer is dead. The cougars have enough meat to feed them for several days.

CLOSE-UP

Scream

Cougars are quiet animals, but sometimes they scream. A cougar's scream has been compared to a female human's scream. Female cougars scream to indicate when they are ready to mate, and males scream when fighting over a female.

3 YEARS

- Males reach sexual maturity
- Males return to range after mating
- Females live alone when not rearing cubs

There is another threat to cougars. For years, researchers believed cougars in the Yellowstone National Park area were dying of starvation. Instead, they learned that some of these cougars had died from **plague**. This contagious disease can be passed to cougars by flea bites or by eating animals that have been infected by plague. In 2020, it was reported that nearly 43 percent of cougars studied in Wyoming tested positive for plague. Four of these cougars died from the disease.

CLOSE-UP

Retractable claws

Like all cats, cougars have retractable claws. Retractable claws can be pulled in when not needed. When the claws are needed, the cougar can quickly produce them. These claws help when climbing trees and catching prey.

FEATURED FAMILY

Practice Makes Perfect

The pair of cougars has now separated. The two adults travel by themselves, always on the lookout for food. One has become a skilled hunter and rarely eats rabbits or squirrels now. His favorite meal is deer, and he has learned how to keep his food supply stocked. The cougar is an expert at killing deer swiftly. After eating a few chunks of meat from a fresh kill, he buries the deer to keep it away from other animals. Then he returns for several days to finish his meal.

ADULT COUGARS

live alone within a set range of land.

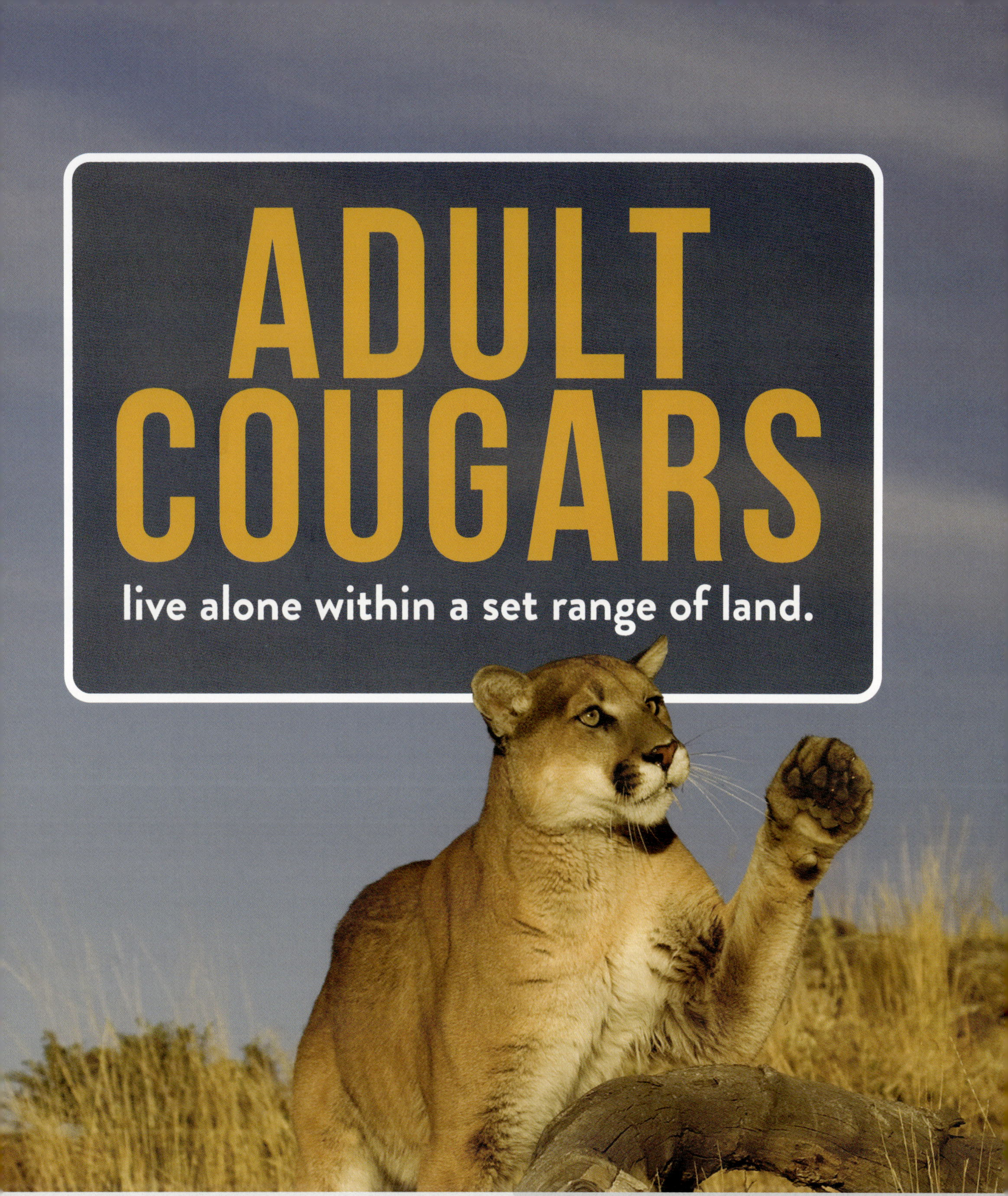

4 YEARS

- Females have one litter every two to three years
- Males mate with multiple females

10 YEARS

- End of life

CHAPTER FOUR

HELPING COUGARS SURVIVE

Cougars easily adapt to new surroundings. This adaptability is why they survive in such different environments, from the Yukon Territory in Canada to the southern tip of South America. In the 1500s, cougars could be found across the United States. Their U.S. territory has since shrunk to mainly the western half of the country with an estimated population of 20,000 to 40,000.

Hunting is a constant threat to cougars. In the early 20th century, farmers and ranchers blamed cougars for killing livestock. As a result, many cougars were hunted and killed. In 1920, the cougar population in California dropped to 600. The state has since banned recreational hunting, but cougars are still killed when people feel threatened by them. It is legal to hunt cougars in other states, as well as in some areas of Canada, Mexico, and South America. Approximately 3,000 cougars are killed by hunters annually in the United States alone.

In general, cougars are not an endangered species. Only the Florida panther subspecies is considered endangered, and it is illegal to hunt them.

Becoming roadkill is another danger for cougars. In California, about two cougars die each week after being hit by a car.

Cougar attacks on humans are rare. Since 1868, there have been only 29 fatal attacks on humans by cougars in North America. Cougars will try to keep away from humans in the wild. Most attacks have happened when people were in the cougar's territory. It is difficult to know where a cougar's territory is. Signs of a territory include mounds of earth, pine needles, and leaves soaked with cougar urine and feces. Younger cougars are more likely to attack humans, but such attacks are rarely fatal. Like all cats, cougars like to chase moving objects. If a cougar is seen in the wilderness, it is best not to run. Look big and keep small children and pets close. Cougars do not like to be cornered. Give cougars a way out and they will most likely take it.

Cougars are a part of the North American landscape. While they once roamed the continent, their territory has shrunk over the past 200 years. This silent, stealthy hunter deserves to reclaim its place in the wilderness and in the world.

FAMILY ALBUM

SNAPSHOTS

The only cougar to live east of the Mississippi River is the **Florida panther.** Most Florida panthers live in southwest Florida. It is estimated that there are about 200 adult panthers left.

Eastern cougars are a subspecies of cougar now believed to be extinct. They were found in the northeastern United States up until the late 1800s.

Some people still claim to have spotted Eastern cougars, but it is more likely that they saw another subspecies of cougar, such as the **North American cougar**.

South American cougars dine on vicuña and guanaco, animals that are both closely related to llamas and alpacas. Vicuña makes up 80 percent of a South American cougar's diet.

In the Andes Mountains of South America, the cougar is regarded as a snatcher of souls or a helper of people.

The **Patagonia puma** is the largest land carnivore in Chile.

While most cougars have pale tan coats, some coats on **Costa Rican cougars** are a reddish-brown color.

The largest populations of Costa Rican cougars are found in national parks such as Santa Rosa and Corcovado, where they are protected from hunting and poachers.

The oldest cougar fossil records date back to 300,000 B.C.

The cougar is one of six species of wild cats found in North America. Unlike other big cats, they can't roar.

With more than 40 different names, cougars hold the Guinness World Record for the animal with the highest number of names. Common names include puma, mountain lion, catamount, panther, painter, American lion, deer tiger, Long Tail, and Night Screamer.

WORDS to Know

adapt	to change to improve the chances of survival
ambush	a suprise attack
carnivore	a meat-eating mammal
dewclaw	an inner toe on a mammal's paws that helps with climbing and catching prey
elusive	hard to find or capture
encroachment	the gradual taking over of an area
over-browsing	the excessive eating of plants and trees, leading to damage
plague	a contagious bacterial disease
retractable	able to draw back or in
subspecies	a category of living things with shared characteristics that ranks below a species
territorial	relating to the owner of an area of land
uniform	consistent, unchanging

LEARN MORE

Books

Geister-Jones, Sophie. *Cougars*. Mendota Heights, Minn.: Apex, 2022.

Hogan, Christa C. *Mountain Lions*. Lake Elmo, Minn.: North Star Editions, 2017.

Riggs, Kate. *Cougars*. Mankato, Minn.: Creative Education and Creative Paperbacks, 2022.

Websites

"Cougar." USDA Forest Service.
https://www.fs.usda.gov/detail/npnht/learningcenter/kids/?cidlfseprd887643

"Cougar or Mountain Lion." Washington NatureMapping Program.
http://naturemappingfoundation.org/natmap/facts/cougar_k6.html

"Mountain Lion." National Geographic Kids.
https://kids.nationalgeographic.com/animals/mammals/facts/mountain-lion

Documentaries

Anderson, Casey, dir. *The Mountain Lion and Me*. Bozeman, Mont.: VisionHawk Films, 2018.

Johnson, David, dir. *Cougars Undercover*. London: BBC, 2015.

Ruggiero, Colin, dir. *Tracking Notes: The Secret World of Mountain Lions*. Missoula, Mont.: Rhythm Productions, 2023

Note: Every effort has been made to ensure that any websites listed above were active at the time of publication. However, because of the nature of the Internet, it is impossible to guarantee that these sites will remain active indefinitely or that their contents will not be altered.

Visit

COUGAR MOUNTAIN ZOO

With a focus on endangered species, this zoo teaches visitors about many kinds of animals, including cougars.

19525 Southeast 54th Street
Issaquah, WA 98027

NORTH CAROLINA ZOO

Meet a pair of cougar siblings! Heath and Olive have been at the zoo since 2014 and love to play, eat, and sleep.

4401 Zoo Parkway
Asheboro, NC 27205

MONTEREY ZOO

This zoo is home to several cougars, including one with a former career in television and movies.

400 River Road
Salinas, CA 93908

INDEX